Fish
men
fear...
SHARK!

BY JERRY & IDAZ GREENBERG

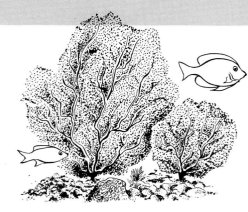

First Edition
1st Printing, January 1969
2nd Printing, January 1970

Published by
SEAHAWK PRESS
6840 S.W. 92nd St.
Miami, Florida 33156

PRINTED IN U. S. A.

Contents

Foreword

All sharks are carnivorous, and most sharks are potentially dangerous to man. Despite this, we had no name for these deadly creatures until 300 years ago. Sailors first used the Spanish name tiburon, and our word "shark" came later, probably from the German word schurke, meaning crafty rogue. The one universal truth that can be applied to sharks is that they are, like man, unpredictable. Certain behavioral patterns emerge, but there is no explanation for the erratic individual.

There are over 225 species, ranging from sharks that mature at less than 10″, to the largest of all fishes, the whale shark, known to reach a length of at least 45 feet. While the whale and basking shark, another large shark species, feed on plankton and therefore pose no real threat to man, all other sharks must remain suspect.

This book is intended as a thumbnail guide to the major species of dangerous sharks found in the waters of Florida and the Caribbean. It will draw from the personal experiences of the author-photographer in over 20 years of observing the creatures of the deep. All photographs were taken in the open seas and coral reefs, unless otherwise noted.

The sea is the last unconquered inner space left to us, and since we will become more and more dependent on it for its bounty of food, minerals, and living space, man must learn more about the dangerous denizens of this last frontier.

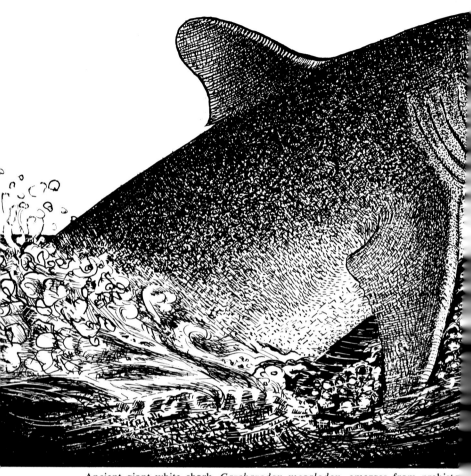

Ancient giant white shark, *Carcharodon megalodon,* emerges from prehisto

Living fossils

The shark is an incredibly efficient and well adapted machine The modern shark is a descendent of primitive sharks known to have existed 340 million years ago. The age of man, in contrast to this, is only about two million years. The giant white shark, *Carcharodon megalodon,* grew to a length of 40 feet or mor in the Miocene era of prehistory. Though smaller, the presen

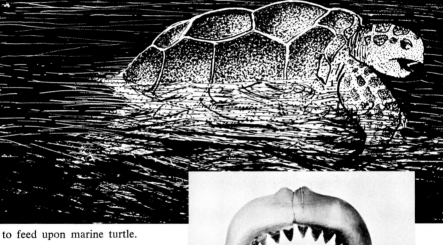

to feed upon marine turtle.

Reconstructed jaw of this mammoth species showing relative size of man.

...ay great white shark closely ...esembles this ancient species. ...he shark dominated all other ...orms of life during the age ...f fishes, and has yet to be ...onquered by man.

4

Susan, Michael and Mimi Green-
berg have room to spare inside the
jaw of a sixteen foot tiger shark.

Sea wolves

Voracious cannibal of the d

Carcharhinidae is the largest family of sharks, with over 60
species. They are sometimes called requiem sharks, after the
first word of the mass for departed souls. They have earned this
name, as they include the bulk of sharks dangerous to man.

The largest and most notorious member of this family is the
tiger shark. It eats anything, including such natural foods as fish,
lobster and stingrays, and such unnatural things as tin cans, coal,
garbage, and people. At least two tiger sharks found in the waters
near Florida had human remains in them, but it is not known if the
victims were alive or dead when eaten. The tiger is well named: it
is strikingly marked with vertical stripes along the sides, which

ger shark, *Galeocerdo cuvieri,* found in all tropical and sub-tropical waters.

fade somewhat with age, and it can easily match its jungle namesake in ferocity of appearance. The tiger grows to a length of about 14 feet in our waters, but is reputed to attain a possible length of 30 feet. One of these blunt-bodied sharks weighed 1300 pounds at 14 feet. Tiger shark teeth are easily distinguished from those of other sharks by their saw-toothed formation, with a deep notch on one side. These serrate teeth, along with the powerful jaws, are capable of cutting pieces from large sea turtles.

All the members of the carcharhinid family bear live young, with an average of 4 to 50 pups per litter. As is true of all sharks, fertilization is internal.

Gracefully gliding over deep waters, accompanied by banded pilot fish, this stream

Silky shark, *Carcharhinus floridanus,* found in both the Atlantic and Pacific Oceans.

...etip shark, *Carcharhinus longimanus,* is a large species believed dangerous to man.

The whitetip is a "blue water" member of the requiem sharks. The oversized pectoral fins, the large blunt dorsal fin and the tail are all clearly marked with white tips. This shark grows to about 13 feet and is usually solitary.

Another pelagic species is the silky shark, whose name is derived from the comparatively smooth feel and look of its skin. Silkies attain a maximum size of about ten feet, but even when quite small are prone to aggressive behavior. It is amazing that this common, easily recognized shark was not identified until 1943.

The whitetips and silkies occur in the same area, and both are considered dangerous, but in my experience the silky is by far the more aggressive of the two. They travel in groups, and have often come in with no hesitation to "check" on me, nipping at my fins. My shark billy discouraged them from further encroachment.

The bull shark, also known as the cub or ground shark, is very commonly seen around bridges and piers. Bulls adapt easily to brackish water, and have penetrated vast distances into fresh water, sometimes to bear young.

Another shark that pokes its way into rivers is the lemon shark, so named for its yellow-brown coloring when young. Though the coloring fades with age, the shark can be recognized by the fact that its second dorsal fin is almost the same size as its first.

The lemon and bull sharks do not travel together, but are both inshore species, and are found in the same areas. Both species reach a length of about 11 feet, but the lemon is slim bodied, and the bull quite a bit heavier and more compact. Other sharks in this same large group of carcharhinids are the brown shark, *C. milberti,* the dusky shark, *C. obscurus,* the blue, *Prionace glauca* and the blacktips. There are two Atlantic blacktip sharks, the small, *C. limbatus,* and the large blacktip or spinner shark, *C. maculipinnis.* The spinner grows to about 8 feet, and is known for its habit of leaping out of the water and doing somersaults in midair.

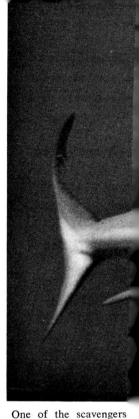

One of the scavengers

Unlike most sharks wh

sea, the bull shark, *Carcharhinus leucas,* will eat almost any offal tossed overboard.

m or die, the lemon, *Negaprion brevirostris,* can rest motionless on the sea bottom.

Ocean rovers

The great white shark, also called man-eater, is not found in abundance anywhere, yet it has been implicated in more attacks upon man than any other shark. Its normal diet includes large mammals such as sea lions, and this swift, powerful, most voracious of sharks is known to attack even boats without provocation.

The young are over 100 pounds at birth, and speculation has it that mature great whites may grow as large as 40 feet. They are not uncommon at 20 feet, and can weigh 6000 pounds at that length. The great white and the mako are members of the Isuridae, or mackerel shark family. The mako is much smaller and lighter bodied than the great white, but both can be recognized by their lunate tail and pointed snout. The mako is the only shark swift enough to feed on swordfish and mackerel, and the only shark listed as a salt water sportsman's game fish.

The most savage of all maneaters is the great white, *Carcharodon carcharias.*

Swiftest of all sharks, *Isurus oxyrhinchus,* the mako. Leaping spectacularly when hooked, this fierce fighter is a favorite of sportsmen.

Hammerhead

The family of sharks Sphyrnidae, commonly called hammerhead, is the most readily identified of all groups. For no proven function, the eyes and nostrils are located at the tips of the lateral lobes of the head. There are two members of this family most often seen in the geographical area in discussion, the common hammerhead, *Sphyrna zygaena,* and the great hammerhead, *Sphyrna mokarran,* shown in this photograph.

There have been several well authenticated attacks on man by these powerful swimmers and omnivorous feeders, and while they are presumed to eat mainly stingrays and other fish, specimens have been found with such diverse objects in their stomachs as beer cans and their own young.

Fellow travelers

Commensals, meaning those who eat at the same table, is the term that describes several species of fish who accompany sharks. It is believed that they benefit from this arrangement by sharing in the smaller morsels of food when the shark feeds, and possibly they also feed on their host's external parasites.

The family Echeneidae includes sharksuckers and remoras. They attach themselves to large fish by means of a suction disc on top of their heads. The pilotfish, rudderfish and cobia are free-swimming commensals, but stick close to their hosts.

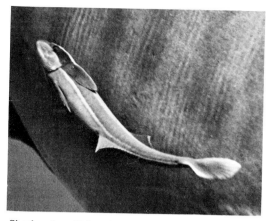

Sharksucker, *Echeneis naucrates,* rides a tiger shark.

Remora, *Remora remora,* favors offshore shark.

Bull shark is host to a remora below it and a sharksucker on its side.

Pilotfish, *Naucrates ductor,* is a member of the jack family, and swims freely above this whitetip shark.

16

This ravenous silky shark attacked with such speed that Don Nelson, with ba

Man vs. shark

The bulk of the sharks dangerous to man, excepting the great white which I have never seen, have served as subjects for my camera. In all my years of photographing these animals, I have never been attacked by any of them. The only one I ever felt I had to destroy was the silky shown on this page, who came in too close after

ck ready, barely had time to push it off. He killed the silky on its next charge.

being hand fed from my boat. It has always been one of my cardinal rules never to let a shark get too close. When shooting an adventure or natural history story, I take the precaution of having one diver with me to "ride shotgun" with a bang stick. I cannot concentrate on taking close-ups of sharks when I must also worry about one nudging me from behind. Experience has taught me that first of all, the menace from sharks is often exaggerated, as sharks are not seen that frequently. Secondly, when sharks are in the area, they are not there necessarily to attack man. They are part of life in the sea.

Aggressive whitetips are held at bay by skin diver. The bang-stick is disarmed here an

Properly armed, man can approach equality with a shark in its domain. Photographs on this page, taken in the shark infested waters of Tongue of the Ocean in the Bahamas, show a diver not only holding his own, but being the aggressor.

Anything that helps man to be more at home in the water, such as mask, snorkel, shark billy and fins, will definitely bring him to a more equal footing with the shark. It is my feeling that these should be standard equipment for survival at sea. A person so supplied has a much better survival outlook than a man with just a life jacket.

being used as a shark billy.

Cage that Jack Stemples built of light-weight steel mesh
is easily launched from boat. This is the only safe way
to observe and photograph sharks, but the author foregoes
this protection in favor of the greater mobility outside.

Sharks
in
captivity

Captain Emil Hanson, pictured at the helm of the Miami Seaquarium 65-foot steel hulled boat, goes on shark hunting expeditions twice a year. The supply of sharks in the huge, glass-windowed tanks must be replenished frequently. The average life span of the shark in confinement is only about one year. As a sideline to stocking the tanks, Capt. Hanson also provides shark parts, such as hearts, livers, and digestive tracts, for scientific research.

In the course of his career, he has learned a great deal about his paradoxical quarry. Sharks have survived hundreds of millions of years of change and upheaval in the world, and yet they are extremely delicate when captured. They frequently perish from shock right on the lines, showing no signs of physical damage.

Capt. Hanson has found that the faster and more active the species of shark, the lower the survival quotient when hooked. This may be because, unlike the bony fishes, they have no air-filled bladders to give them buoyancy. They either swim or sink. Only in movement does oxygen-bearing water pass over their gills, and the hooks may inhibit this process for most species.

Tiger shark eagerly approaches a hook. Bait used is fish or whale meat.

To compound the difficulty in capturing sharks alive, the free swimming sharks in the same area with hooked sharks will often dine on their luckless brothers. The worst offender in this respect is the tiger shark, which often hooks itself while dining on another shark. Hammerheads seem to appeal particularly to the cannibal instinct in other sharks and are seldom taken alive.

Crew of two pitches in to boat day's catch.

Last burst of energy fails to disgorge hook.

Tiger in a tank, one of the pri

Once in captivity, a further difficulty arises in that most sharks will not feed. While the fat in their liver, which comprises about one-fourth their total body weight, can support them for weeks between feedings, they are still doomed to die. One tiger shark, after refusing food when captured and put into the tanks, moved about constantly until its death four months later.

actions at the Seaquarium, views visitors with disinterest from his new home.

Graphic
section

These sharks, after an exploratory circle or two, are grabbing the dead bait fish. From their position, one can see why the myth of a shark having to roll over on its belly in order to bite has been disproven.

In cases where the bait was alive and still struggling, the sharks came in with little hesitation and immediately attacked. One hammerhead was in such haste to devour the meal that it grabbed the fish and impaled itself upon the spear. Undaunted, it continued the meal and then, spear still in its mouth and sticking out through the gills, swam off.

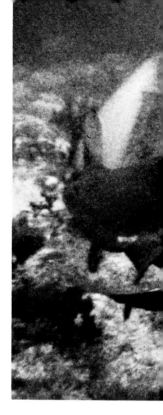

Top, lemon, left, bull, right, hammerhead

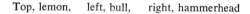

courtesy National Geographic Society ©1962

Whitetip sharks

Bull sharks

31

Whitetip sharks

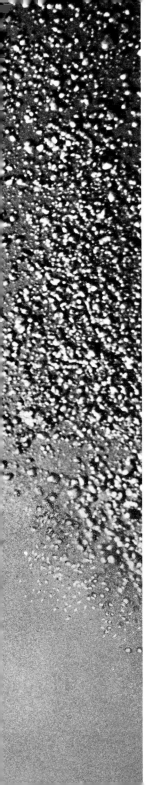

When the diver becomes confident of his ability to cope with the shark in its own environment, some of the fear he has felt disappears, and is supplanted by admiration for the grace and vitality of the animal. Man must still respect the danger posed by these wild, unpredictable lords of the sea, however, and never turn his back to them.

Shark research

Scientists are engaged in a two-pronged attack on the shark, the first of which is the study of shark behavior. The Armed Forces during World War II were concerned over the fear servicemen had of shark attack. In order to allay these fears and aid in survival at sea, the government ordered development of a shark repellent. It was believed that the smell of decomposed shark and a dense cover screen would repel attack. From this came a chemical compound of copper acetate and nigrosine-type dye. This proved to be largely ineffective for anything but morale, mainly because all sharks do not react the same way to anything. The repellent did point up our need for more extensive knowledge of shark behavior.

The second phase of shark research is more recent. Sharks seem to have a natural immunity against cancer, heart disease and other ailments that man is prone to, and scientists hope to unlock the secret of that immunity in order to serve man.

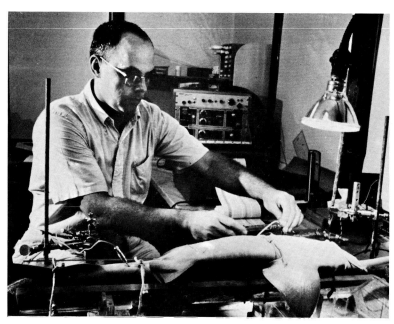

Dr. Frederick Sudak does circulatory research on an anaesthetized nurse shark.

Dr. Arthur Myrberg operating the television console at the Lerner Marine Laboratory on Bimini Island.

Cleaning day for the underwater housing of this television system which enables scientists to study sharks in their natural habitat.

Sound research

Dr. Donald Nelson obtains an electro-cardiogram of a trained shark, whose he

A t the University of Miami's Institute of Marine Science, some of the research is aimed at finding out just what the shark sees, hears and smells. Here, in noise-controlled chambers, sharks are being conditioned to respond to sound by means of an electrical shock. Research has shown that lemon sharks, used because they are hardy and survive best in captivity, can hear only low frequency vibrations from 10 to about 800 cycles per second. This is within

involuntarily skip a beat when it hears a sound previously associated with shock.

the range of sounds made by struggling or wounded fish. Proof that sharks are attracted by these sounds was obtained when recordings of these vibrations were made and played underwater. These recordings drew sharks from as far away as 900 feet. The sharks home in on these sounds by using their extremely sensitive vibration detectors, the inner ear and lateral line system. The latter is a highly developed network extending along the body and face.

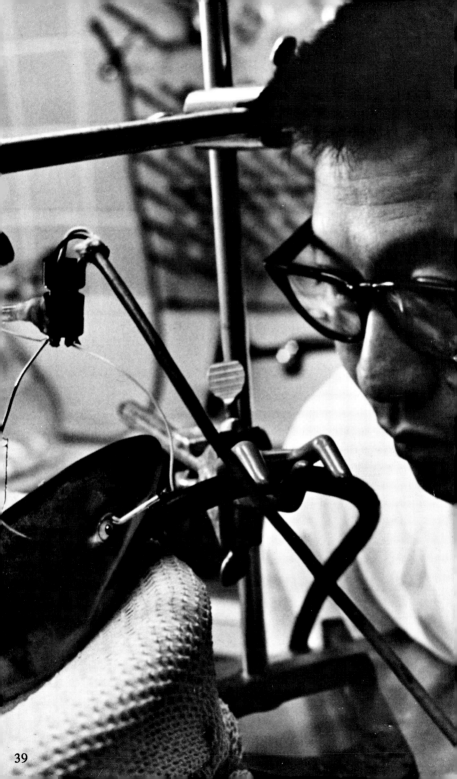

Sight and Smell research

S harks are able to see in very dim light situations. This is possible because the shark eye has thousands of little mirrors, which reflect the existing light back onto the retina.

The shark eye also has a third eyelid, or nictitans, which moves freely to cover the exposed portion of the eye. It was formerly thought that this nictitans was used by the shark to shield the extremely light-sensitive eyes in bright light, but it has since been established that it is used to protect the eye from injury.

Instead, when the light level is too bright for the shark, pigment cells send down "shades" to cover the mirrors. In addition, the shark eye has a pupil which dilates and contracts with light, unlike other fish. Because of these mechanisms, it is now believed that

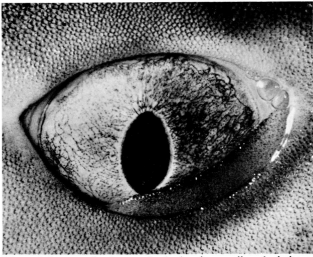

View of the eye of a lemon shark showing pupil and nictitans.

Dr. Duco Hamasaki studying electrical activity of the shark's retina.

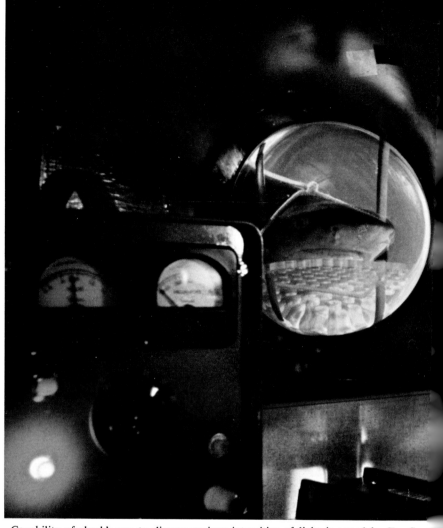

Capability of shark's eye to discern various intensities of light is tested by Dr. San

sharks are not the exclusively nocturnal animals they were thought to be, but rather that they are able to feed day and night, at will. They are far-sighted, and therefore have poor ability to distinguish the details of an object, but can readily differentiate a moving fish from its background.

It is certain that sharks rely on their vision when close to prey, but beyond their visual limits, their sense of smell is dominant. The brain of the shark has been called a "brain of smell",

er. Small lemon shark perceives light beam flashed at it through plexiglass bubble.

as the larger part of it is given over to olfactory sensations. Sharks can follow an olfactory corridor even when the prey fish is not visible. They need not depend on the smell of blood, as fish in distress apparently give off a secretion that attracts them.

It is not known whether sharks can see color. If they can, it is conceivable that a color or pattern could be found that is repellent to them. Combined with an equally offensive sound or smell, perhaps an adequate protection for man could be devised.

Dr. Sylvia Mead does comparative studies of shark teeth at the Mote Marine Labc.

Bite research

The most certain way to identify any shark is by its teeth. They are usually pointed, sharp, and serrate, excellent for holding and cutting. Sharks have 5 to 15 rows of teeth in their mouth, which are constantly moving forward and replacing missing teeth. In addition, the skin is plated with tooth-like denticles.

Fossil records of early sharks were left mainly by their teeth. Sharks are elasmobranchs, with a skeletal structure made up entirely of cartilege, which is much softer than bone and does not usually leave a fossil record. Generally, when the shark is feeding, it grabs its prey between its powerful jaws, and uses a fierce sawing motion to bite off massive chunks.

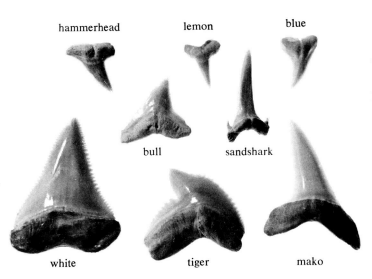

hammerhead   lemon   blue

bull   sandshark

white   tiger   mako

tooth specimens courtesy Miami Seaquarium & Al Pflueger, Inc.

Tooth specimens above are shown actual size.

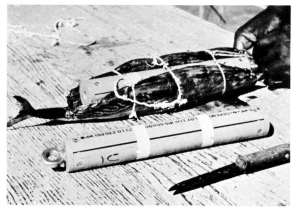

Sandwich for a shark. Bait wrapped around
"bite-meter" measures power of shark's bite.

Dusky shark seizes inedible meal and records bite.

Force of dusky's bit
estimated by Dr. Perr
Gilbert at about 18
per square inch. Dr.
bert is now directo
Mote Marine Laborat

Tests made in the Lerner Marine Laboratory utilized a unique device to determine the power of a shark's bite. This was a "bite-meter", consisting of an aluminum core of specific hardness in a specially designed steel casing, which was wrapped in a bait fish. When the shark bit this, the force of the bite dented the aluminum core, and the tons of pressure per square inch were recorded.

Sharks do not rely only on biting to obtain food. They have the ability to open their jaws so wide that they seem dislocated, and can swallow large prey in one gulp. A tiger shark weighing 900 lbs. was found to have consumed, whole, a 200 lb. hammerhead.

One of the first rules of safety in shark-infested waters is to leave the wate

RULES TO FOLLOW IN TROPICAL WATERS

Shark
sense

- Don't ever swim or skin dive alone.

- Stay out of the water if dangerous sharks are known to be in the area.

- Don't swim with a bleeding wound. Boat all speared fish immediately.

- Do not provoke or molest any shark or spear one, no matter how small.

as a shark is sighted. A shark cruises and is observed safely by diver in boat.

● If you sight a shark, don't panic. Get out of the water as soon as possible, swimming with a smooth, rhythmic stroke. If a shark moves in, hit it on the snout with the heaviest instrument available. Use your fist only as a last resort, for the shark's skin is rough and will cause bleeding.

● Remember that the odds of being attacked by a shark are less than the risk of death by lightning or drowning.

These rules are based on information from the Shark Research Panel, supported by the U.S. Navy, which gathers shark attack data from all over the world. It consists of such experts as the Drs. Perry Gilbert, Leonard Schultz, Albert Tester, and Stewart Springer.

Other publications by Seahawk Press

UNDERWATER PHOTOGRAPHY SIMPLIFIED $2.00
ADVENTURES OF A REEFCOMBER $2.00
CORAL REEF STATE PARK COLOR MAP $2.00

Published by
SEAHAWK PRESS
6840 S.W. 92nd St.
Miami, Florida 33156
LITHO IN U.S.A.
MIAMI POST PUBLISHING CO.